the backyard dairy book

by len street

and andrew singer

Copyright 1975: Len Street and Andrew Singer

Originally published in 1972 by Whole Earth Tools,
Andrew Singer's own 'backyard' publishing venture.

This new, greatly extended, edition is published by:

PRISM PRESS
Stable Court,
Chalmington,
Dorchester, Dorset DT2 OHB.

and printed by:

UNWIN BROTHERS LIMITED,
The Gresham Press,
Old Woking,
Surrey.

SBN 0 904727 05 X Cloth edition
SBN 0 904727 06 8 Paperback edition

contents

4

about this book

We have written this book, without apology, as propaganda.
Its aim is to provide enough basic information to encourage
you, the reader, wherever you are, to start your own
dairy production. More and more people are getting interested in reducing their reliance on centralised factory food
production by growing their own vegetables in a garden or
allotment; planting fruit trees and bushes; keeping a few
chickens or having a go at beekeeping. An important ele-
ment in any move towards increased self-reliance is home
dairy production - so read on.

This book claims only to be an introduction. It tells you why
it is worth starting dairy production; what you will need;
what will be involved and how you can expect to benefit.
Once it has done its job - of persuading you to take the first
step - its function is finished. It then refers you to more
comprehensive works for you to use as you get deeper into
the subjects concerned.

We hope that you think this book is worth buying, both now
and more especially after you have finished reading it. But
even more, we hope that it achieves its stated purpose -
it spurs you to action.

Have fun.....

Len Street and Andrew Singer.

about the authors

LEN STREET lives with his wife June in a hidden valley just outside St. Blazey in Cornwall. Having already built one very beautiful Cornish cottage out of local stone and slate, they are in the process of rebuilding another cottage in a similar style. For a living, Len makes pottery wall-plaques printed from old farm butterprints and also working pottery butterprints for home use. He has also produced a kitchen wall-chart on butter making at 45p including postage. Details of all these are available from him at Wheal Cottage, Prideaux Road, St. Blazey, Cornwall.

June and Len Street have been homesteading in Cornwall for eight years and have had considerable experience keeping both cows and goats, and also, of home butter and cheese making. They were recently featured on BBC TV's 'Man Alive' on the subject of self-sufficiency homesteading.

ANDREW SINGER, until recently, operated with his wife Glo a family publishing venture known as 'Whole Earth Tools'. They did their own design, typesetting, printing, binding, selling and distribution as a way of proving that even 'mass communications' could be decentralised, even down to the smallest of units. The first edition of this book was originally published by 'Whole Earth Tools'.

Andrew and Glo Singer live with their two sons in Cambridgeshire, where they have homesteaded for three years. They have had experience of keeping a house goat. Andrew's next book will deal with backyard egg production to be published by Prism Press in 1976.

Len provided most of the practical experience and Andrew supplemented it with all the necessary research.

acknowledgements

The authors would like to thank the following for their help in the production of this book:

> Mr. and Mrs. B. Franche of Martock, Somerset who are champion goat keepers.
> Robert and Brenda Vale of Witcham, Cambridge who are ex-goat keepers and have a house cow.
> Mr. R. Wilson of Fairford, Gloucestershire who keeps Dexter cows.

The following readers of the first edition provided useful information which has been included in this edition:

> Jim Platts of Willingham, Cambridgeshire.
> Richard and Joan Collins of Westbury, Somerset.
> H. Christian of Woodseaves, Staffordshire.
> Chris Bennett of Ely, Cambridgeshire.

Many other readers of the first edition wrote in with useful criticisms and comments, for which many thanks. We are hoping to receive more on this edition. Write to Len or Andrew, care of Prism Press, and it will be forwarded. If you want a reply, please make sure that you enclose a stamped and addressed envelope.

1 why we want to persuade you to start dairy production

At the very least, home dairy production will involve you in an outlay of £25 and half an hour of work per day, every day. It will more probably involve you in quite a lot more. So, why should you bother? We will give you some good reasons.

CASH SAVING

The less CASH you need to buy the necessities of life, the less you need to earn by selling your time as an employee, or by selling the products of your work as a self-employed person. IF YOU ARE PAYING TAX AT THE STANDARD RATE, EVERY POUNDSWORTH OF FOOD WHICH YOU PRODUCE YOURSELF SAVES YOU HAVING TO EARN £1.43p. Remember that.... Furthermore, if you can keep your cash needs below £12 per week, then you should be able to get away without having to pay National Insurance stamps, so saving yourself as much as £2.41 per week.

So, if you are interested in reducing your income needs, where do you start? Food and drink are obviously the items on which most families spend the bulk of their income. In fact, according to the latest Family Expenditure Survey in 1973, the average UK household spent 29% of its total expenditure on food and drink - £11.58 at <u>1973</u> prices. This was split down as follows:

	£
Milk	0.71
Cheese	0.20
Butter	0.18
Other Dairy Products or Dairy Substitutes (ice-cream, tinned milk, cooking fats)	0.32
Beef and Veal	0.57

TOTAL FOOD EXPENDITURE WHICH COULD BE HOME-PRODUCED IN A BACKYARD DAIRY : £1.98

Other items in the food budget, for comparison, were:

	£
Vegetables	0.84
Fruit	0.48
Eggs	0.29
Poultry	0.91
Meat (other than beef, veal and poultry)	0.96

So, growing your own vegetables will save you money, and keeping chickens will save you money, but if your consumption patterns are close to Mr. and Mrs. Average Family, a backyard dairy could save you most of all. If you spend less than average (as we might suspect) on items such as 'Meals Bought Out' (£1.41), 'Sugar, Sweets and Jams' (£0.46), and 'Alcoholic Drink' (£1.85), then you probably spend more than average on the products you could produce in a backyard dairy. At today's prices, a backyard dairy could save you around £3-4 in bought-in food - in other words, it could theoretically reduce your before-tax income needs by £4-6 per week.

ENJOYMENT AND SATISFACTION

For many people, this is the prime motivation in attempting
home food production. There is certainly a great deal of
pleasure and satisfaction to be had from keeping a house-
cow or house-goat. When properly reared, either type of
animal becomes an affectionate family pet - a member of
the family. Then there is the enormous satisfaction of
cheese and butter making. Both demand a high degree of
personal judgement and care, but in both cases the home
producer can expect to produce products greatly superior
to those from the factory. Cheese making in particular,
offers so many variations in technique, and many different
flavours and consistencies that it can become an absorbing
hobby in itself. Like wine making, it has the particular
fascination of controlling and utilising a natural live process.

INDEPENDENCE FROM THE SYSTEM

This may also motivate you to start home dairy production.
We can all sit on our backsides and moan about the ecological effects of modern farming methods, about the power
of modern industrial corporations and about people having
to spend their lives as factory automatons. Unless you do
what you can to cut down your economic reliance on this
situation, you are as much to blame for what you are
moaning about as anyone. Now that you have picked up
this book, there is no excuse. We are going to tell you how
you can reduce this reliance considerably. Combined with
a switch to vegetarian eating, and the cultivation of a small
vegetable and fruit garden or allotment, home dairy production could make you 80-90% independent of the system in
foods. (Look out for Andrew Singer's book on backyard
egg production coming early in 1976).

The milk you are buying right now is probably around three
days old; and it has been pasteurised (heated to destroy
bacteria). It may also have been left out in the sun and thus
had some of its vitamins destroyed. It is probably produced

11

by cows in large commercial herds, pushed by careful feeding of pre-mixed factory foods to their limits of milk production. Treated this way, cows are far more likely to suffer from mastitis and other udder troubles. They have to be injected in the udder with penicillin or other antibiotics. In a recent study of 40,000 samples of milk, 11% contained penicillin and 1% other antibiotics. This creates a danger of immunity to these drugs in humans and also of possible direct harmful effects on the body. One antibiotic used to treat mastitis, chloramphenical, can fatally disorganise the marrow of the bone. Surely, fresh milk from your own animal must be better for you than what you are getting now.

In addition, there may still be danger from the chlorinated hydrocarbon group of pesticides (DDT, Aldrin, Dieldrin etc.) which build up to high concentrations in the fat of animals grazing on treated land and can result in severe contamination of the fat in milk (See Reference 2).

Dairy farming, like all branches of farming, has been concentrated into large units by agribusiness. The US now boasts huge 'cowtels' which have brought their own pollution problems. One of these super-farms can produce as much shit as a fair-sized town. And as you can guess, it is not 'economic' to actually use this magnificent natural fertilizer so they have been known to channel it for direct untreated dumping into nearby rivers, thus throwing out the ecological balance of large river systems.

Under the UK Public Health (Food Preservatives) Regulations, 1925-1940, the addition of preservatives is officially prohibited in all dairy products. However, there have been investigations in the USA and elsewhere of hydrogen peroxide being used to 'improve' the quality of milk for cheese making. That's the stuff that you can use to dye your hair blond - we wonder what it does to your stomach.... (See Reference 3).

12

Apart from the pure health aspects, you may be spurred on faster to home dairy production by a little knowledge of the kind of places that now do the work previously done by pretty milkmaids in cool dairies. Butter is now widely made by a continous process in large factories, which extrudes it out to be cut and packaged. Some cheese production, notably Swiss and Cheddar, has 'progressed' to the continuous production line stage and you can be sure that they are working on those that can't yet. But none go quite as far as processed cheese, defined as:

> 'The clean, sound, pasteurised product made by cominuting and blending, with the aid of heat and water, with or without the addition of salt, one or more lots of cheese into a HOMOGENEOUS PLASTIC MASS.' (Also Reference 3 - our capitals).

Process cheese (made in slices by organisations such as Kraft) is treated with 10-20% of a special starter which inhibits certain bacteria naturally present in milk but which cause 'gas-blowing' and reduce the keeping qualities of neatly-packaged family sized cheese portions. The antibiotic used, nisin, is said not to have any harmful effects on the flora of the human stomach. Kraft slices are made by rolling presses and the extruded sheet formed is then sliced into strips, which are fed to the cutting and packing machines.

You may also be interested to know that experiments are under way on irradiating cheese with powerful X-, gamma- and other rays to prevent mould growth under plastic wrappers. So, before the dairy technologists get too carried away on their search for the perfectly uniform, infinite shelf-life dairy product, perhaps the time has come to point a rude gesture in their direction and become independent of the whole unpleasant scene.

CHURNING BY DOG-POWER.

2 the basic economics of home dairy production

If you want to learn more about becoming more independent in dairy products, the first shock we want to get you over is that there is no half-way house. With cereal products, for instance, you can save an enormous proportion of bread costs by buying wheat in bulk and grinding and baking at home - without having to get involved in the actual growing of the grain. The same is not true of dairy products. You need around ten pints of milk to make one pound of butter. So you can see very quickly that buying retail milk for butter making is totally uneconomic, certainly at the present subsidised prices we pay for butter in the UK.

The picture is not so clear-cut with cheese. You need about eight pints of milk to make one pound of hard cheese, so that at present UK prices, it could just be worthwhile making your won cheese from bought retail milk. Some of you may feel like attempting to find a source of milk sold cheap because it is slightly off and so get into cheese making that way, but the point we want to make is that, generally, home dairy production only really becomes worthwhile when you have your own milking animal.

Now the idea of keeping a cow or a goat in a London back

The interior of Holland Park Dairy

and

The floor plan of the same Dairy

garden, or wherever you are right now, may seem crazy to you, but it is not that crazy. A goat kept in Bermondsey (in London's dockland) during the Blitz actually won a world milk production prize. In times gone by, even quite small English households kept a cow, and in some districts the labourer's cow was part of his wages. Cows were widely kept in London until a few years ago. The last London herd was at Peckham Rye - until 1959. Rumour has it that there is still a dairyman operating in Liverpool, keeping his cows behind the shop. If any reader knows more about him please let us know.

INTERIOR OF IMPROVED LONDON COW-SHED.

So, if you live in anything at least as rural as a ground floor London flat, keep reading. We have information here which could get you the independence and savings which we mentioned earlier.

3 what animal to start with

All mammals produce milk, but only certain ones are really suitable as regular milking animals - namely the larger domesticated breeds, the cow, the horse, the goat and the sheep. Unless you want to go deeply into self-sufficiency and double up your milking mare as a low cost form of transport we suggest you forget the horse. Milking breeds of sheep exist in France, where sheep's milk is used to make such prize cheeses as Roquefort, but if you are not desperate for the taste of backyard Roquefort, we suggest you forget sheep as well. This leaves us with cows and goats as the most suitable animals for house milkers. Both are widely kept for this purpose and there are many owners with a strong preference for one or the other. This can be confusing for the would-be backyard dairyman seeking advice, so we will try first to give a reasonably unbiased assessment of the main points in favour of each.

IN FAVOUR OF GOATS

* Goats are delightful animals to keep, they are full of individuality and character.
* They will eat almost anything. They can survive on land

18

where most cows would starve.
* A goat is a more efficient converter of foodstuffs into milk
  For every ten stone bag of dairy cake fed, the average
  goat produces two gallons more milk.(See Reference 4)
* The milk production of a goat is about right for the dairy

product needs of the average family. It needs less space
than a cow; costs less to buy;requires less equipment and
less time to look after.
* Goats' milk is said to be easier to digest. People suffering
  from eczema, asthma and psoriasis are often helped by
  changing from cow to goat milk. (See Reference 5)
* With a small trailer, your house goat can be taken along
  on family trips. A cow requires someone at home every
  day to milk it.
* Goats' milk can be frozen.
* You can sell goats' milk without a licence. In order to sell
  cows' milk, you have to invest a lot of money in officially-
  recognised dairy equipment and facilities.

* You are used to the taste of cows' milk. Goats' milk tastes pretty similar cold, we find, but gives off a 'goaty' smell when heated. (This is due to the presence of short-chain fatty acids in the milk. These, in ewes' milk, are responsible for the 'peppery' taste of Roquefort compared with Stilton. See Reference 6.)
* Most breeds of cow will not destroy young trees and bushes as will goats.
* Surplus bull calves can be sold or raised to produce veal and beef. Goats' meat is not as valuable.
* A well-reared house cow is gentle, affectionate and docile. In comparison, most goats are boisterous and capricious.
* If properly trained, a cow can double her usefulness by pulling the plough for you as well.

* Cows' milk cream settles easily and can be obtained without the use of a mechanical separator. The butter is just

as you would expect it to be. Goats' milk butter can be greasy.

* A cow grazes happily with little trouble to its owner and can be left out day and night, summer and winter. A goat has to have shelter at night and even during the day in case it rains.
* Artificial insemination is universally available for cows, but not for goats. Goats have to be transported to the billy for mating.
* The labour costs per pint of milk are much lower. This, of course, is the reason why virtually all commercial milk production is from cows.

Frankly, our advice is that the animal with which you start will depend upon the circumstances:

1. If you or one of your family suffer from excema, asthma or psoriasis, try a diet excluding all cow-based products (meat and dairy). If it helps, then buy a goat.
2. If you have less than about an acre of decent grazing, you are going to have to bring large quantities of food, either gathered or bought, to a cow. The economics of home milk production then become marginal, so go for a goat. It eats less and is less fussy than a cow.
3. If your land is of poor quality, it will probably support only a goat or a Dexter cow.

In all other circumstances, we recommend a cow rather than a goat as the most suitable house-milker. This is mainly because it is so much less demanding and troublesome, and also gives you 100% acceptable milk and easy cream. A cow is very satisfying and relaxing to keep, but it must be admitted that it is a fairly daunting animal for a complete beginner to manage. Whatever your circumstances, there is a lot to be said for starting off with a goat, even if you intend to progress later to a cow. You can learn how to milk more easily on a goat and you will make less expensive mistakes. Goats also have one unfair advantage over cows for any beginner. There are many more 'amateur' goatkeepers around than cowkeepers. They are enthusiastic, they run lots of local clubs where you can learn a great deal and they are generally very helpful to beginners. So start with a goat and then, unless your circumstances are as 1, 2, or 3 above, move on later to a cow.

4 making a start: what you'll need

We are going to assume that you will follow the advice of the last chapter and start off with a goat, possibly progressing after a couple of year's experience to a cow. If you want to go straight to a cow, read all this selectively and try, if you can, either to start with a calf or get lots of local help - from sympathetic farmers, vets or the like.

WHAT GOAT TO BUY?

Apart from choosing between the various different breeds, the first choice is whether to go for a non-pedigree 'scrub' goat through the small-ad columns of your local newspaper, or to invest in a pedigree from a reputable breeder. A 'scrub' kid or goat may cost as little as five pounds and the advantage is that if you make awful beginner's mistakes and she dies or goes out of milk, then you have not lost much. The disadvantage is that you get attached to an animal that will more than likely turn out to give an indifferent yield and you will end up doing as much work for two pints a day as you would to get eight pints from an animal in which you had invested and risked a bit more money. We suggest that you lay out twenty pounds for a four to six month old female kid, handreared by a breeder recommended by other goatkeepers. That is the cheapest start, but of course, you have to wait another eighteen months or so before you get any milk. A year old pedigree goatling would cost thirty to thirty-five pounds and save you six months of waiting. (Note that the Golden Guernseys are about two-thirds of these prices.)

There is a great advantage in getting a kid that has been handfed from birth. It will respond positively to humans and be much easier to handle.

As to breeds of goats, there are a number to choose from:

THE SAANEN is a Swiss breed, pure white and capable of the highest yields on decent grazing.
BRITISH SAANEN is a mixture of the Saanen with native British stock and is recommended for its docility.
TOGGENBURG is another Swiss breed, brown with white markings. It is smaller than the first two and less high-yielding but it is the best breed for grazing rather than browsing.
BRITISH TOGGENBURG is bigger than the Swiss breed and has a higher milk yield.
BRITISH ALPINE is large, black with white markings and difficult to control.
ANGLO-NUBIAN is heavier than the Swiss breeds and it has floppy ears and a Roman nose. It comes in various colours and gives the creamiest milk, but not in great quantity. The breed is very noisy and nervous.
GOLDEN GUERNSEY is not recognised by the British Goat Society. It is a sort of mini-goat and very suitable for back-yarding. It stands about 25 inches to the shoulder and gives four to five pints a day, eating correspondingly less. It is very affectionate and becomes attached to its owner so only buy when it is very young. It is cheaper to buy than the other breeds but it requires special care in cold winters.
'BRITISH' is the name given by the British Goat Society to

24

pedigree cross-breeds. Some champion milkers are in this category.

Any of these breeds are suitable for the beginner, as long as you recognise their problems, for instance, the nervousness of the Anglo-Nubian or the difficulty of controlling the British Alpine. In certain circumstances, the fact that goats prefer browsing on a variety of plants, shrubs and trees may be a disadvantage. The little Toggenburg is the best breed to adapt to a life of grazing on grass and little else.

Having said all that about breeds, the actual breed you will choose almost certainly will turn out to be less important a factor determining success or failure than the milk yielding qualities of the particular stock from which your goat was bred. Yields vary enormously, from under two pints a day to two gallons a day and more. There are a number of dedicated breeders around the country doing their best to improve the quality of British goat stock. Our advice is that, unless

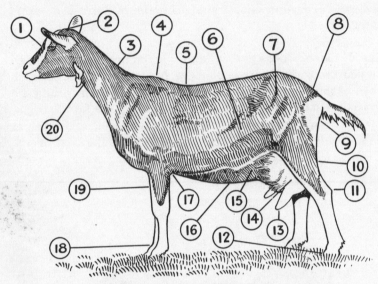

GOOD POINTS IN A MILKER AND BREEDING GOAT

1 - Eye, bright and gentle
2 - Head, shapely and intelligent
3 - Neck, long, not coarse
4 - Shoulders, clean and neat
5 - Back, the line long and level
6 - Ribs, deep and well sprung
7 - Pelvis, wide
8 - Rump, sloping gently
9 - Escutcheon, wide and reaching high
10 - Rear of udder, well developed
11 - Hocks, wide apart and straight
12 - Feet, sound and neat
13 - Teats, pointed and directed forward
14 - Udder, spherical and firmly attached
15 - Barrel, ample for food
16 - Milk veins, prominent
17 - Body, deep allowing room for heart
18 - Pasterns, fairly straight
19 - Forelegs, straight, sound not too close
20 - Throat, clean and fine

you are prepared to treat it as a short-term learning exercise only, pay more and get an animal whose pedigree promises a good milk yield.

In order to locate such an animal, the first step is to write to the British Goat Society at Rougham, Bury St. Edmunds, Suffolk, enclosing a stamped and addressed envelope and asking for the secretary's address for your local goat club. Also, if you have already decided on the breed you want, ask for a list of recommended breeders. Go to a couple of goat club meetings and ask around about breeders and, if you wish, whether any members have good goats in milk or in kid for sale. If you are particularly interested in the Golden Guernsey breed, contact the English Golden Guernsey Goat Club, Henstead Lodge, Harleston, Norfolk.

Judging the likely milking quality of a kid is no matter for a beginner. That is why we emphasise getting local advice through your goat club. The best indicator, if you know how to judge it, is the kid's pedigree, but clearly you want a bright, healthy looking kid with an alert eye. The best goats for milking tend to be non-stop nervous eaters. Lastly, unless you want your goat for chasing scrumping boys out of your orchard, make sure that the kid is hornless or has been dehorned.

HOUSING AND MANAGEMENT SYSTEMS

Depending upon your land and your preferences, there are various systems of management suitable for a house goat.

1. EXTENSIVE

If you are surrounded by open scrubland or moorland where a roaming goat can do little damage, you can let it do just that, providing that there is somewhere for it to shelter when it rains and at night. Your goat will gather a lot of its own food, but reckon on providing concentrates all the year-round and bulk food in the winter. (For an explanation of 'bulk foods' and 'concentrates' see below on feeding.)

2. <u>FENCED</u>

If you have $\frac{1}{3}$ acre or more of decent grazing, your goat will make the best use of it if you fence it into at least two sections, ideally more. Use the sections in rotation. Depending on the acreage, you may have enough grass to make winter hay or silage, which otherwise will have to be bought in or substituted for. On a very small acreage, extra bulk food may also be needed in the summer. Concentrates will be needed all the year. Most breeds of goats like a variety of bulk foods, so you may have to bring in twigs, tree-leaves, garden waste etc. for them - that is why the Toggenburg is so well suited to this system - it is generally happy with just grass.

3. <u>ZERO GRAZING</u>

A new system, labour saving and excellent if you are out all day, and one that produces big milk yields - but you should not keep a single goat this way unless it is in sight of company a lot of the time. The idea is to provide housing with feed-racks and a straw covered sleeping area together with an outside concrete run attached. All food is brought in to the goat, so it wastes no energy in gathering and very little in keeping warm. Much more is therefore converted into milk. Surprisingly, goats thrive under this system and produce splendid yields. It does not sound very labour saving but proponents claim that it is easier to mechanically cut and gather a row of grass every day than it is to maintain and move electric fences in such a way as to make equally efficient use of limited grazing space.

CONFINING YOUR GOAT

The goat is an inquisitive animal, fond of eating all sorts

28

of plants you would rather she did not - like prize rose-
buds about to bloom, or your neighbour's field of kale -
so she usually needs to be confined. You can either tether
or fence her. The trouble with tethering is the labour involved. You have to move her each day to a piece of ground
soft enough to take the tethering stake and hard enough so
that she will not pull it out. Then when it rains you have to
take her in, or move a little portable shelter each day as
well. In our opinion, tethering, though widely practised,
is a lousy way of confining goats. The real way is to fence.
Non-electric fences have to be four foot high to contain a
goat effectively, with posts every six feet. They should be
constructed of sheep fencing, or chain link, but not barbed
wire. Now this kind of fencing has, of late, become enormously expensive, even if you construct it yourself. So
we would only recommend it for zero-grazing exercise
yards or for permanent fencing near to the goat shed. For
other situations, electric fencing is the cheapest solution
and probably the most secure. Its only disadvantage is the
labour involved in maintaining it.

An electric fence for goats consists of three strands - one
at goat-eye height; one just below goat-belly height, and
one half way between the other two. It is cheaper to make
your own posts and buy just the insulators, but follow the
instructions of the fencing power-unit manufacturers.
Most units go 'blip' every second regardless, but modern
electronics plus the inventiveness of a retired dairy farmer
in Lancashire have now made available a unit that only goes
'blip' when the goat touches - clearly, much, much more
economical to run and about the same price to buy. This is
the 'Rossendale' (Rossendale Electric Fencers, Rossendale
Lancs.).

Goats are stubborn and determined animals, once they
decide they would like to get outside their field. To get
your goat to respect the electric fence, you will have to
wait until it is beyond the playful kid stage. Then get a
neighbour or friend to hold its mouth around the fence wire

for three or four good jolts. Unless it is a very stubborn goat, it will not go within a foot of the wire again - ever. Nor will it go near the person who held it - that it why it is not a job to do yourself.

The problem with electric fences is that if the bottom strand touches wet grass, it earths and wastes electricity continuously. Cows conveniently reach under the fence and munch as far as they are able to reach without getting a shock, but goats are too careful to risk that kind of unpleasantness. So you have to go round with a pair of garden shears every so often cutting back potential earthing grass. Some power-units, including the Rossendale, flash lights when earthing is happening, so if you check every day, you need only to cut back the grass when it is actually causing a short circuit.

HOUSING YOUR GOAT

If you are using zero-grazing, you will need a large goat-house with room for the animal to wander round - a minimum of one hundred square feet. Otherwise, if the house is merely to be used as a rain shelter and sleeping quarters a goat can manage with something a quarter the size, but it is convenient to have a shed big enough to do your milking in as well, and to store your feeds, (but in 100% goat-proof part of the shed - over eating can easily kill your goat), so we would advise something about 5ft. x 10ft. at the smallest. If the shed is wood, we suggest that it should be double-walled with some form of insulation between. Storing hay above the goat area if both convenient and insulating for the goat. The shed should be light and well-ventilated.

The floor is easiest if made in concrete. If you lay a damp-course below the concrete, and lay a thin top lay of strong concrete directly over a two inch layer of polystyrene insulation, you can get away without straw bedding. The concrete floor is then warm enough for the goat to lie down on

30

directly. If the floor is made to slope gently to a gully at one wall, which itself drains outside the shed into either a septic tank or a regularly-emptied garden manure tank, then the floor can be swept clean every morning with little trouble. Mucking out straw bedding is a job you do not have to do often, but it is a messy business. Even if you do use straw, though, the slope-plus-gully design helps to keep the straw dry for longer.

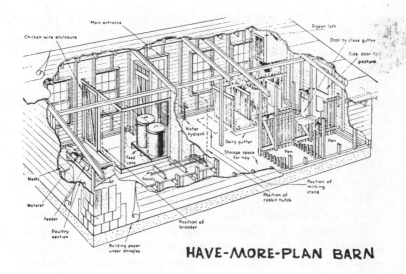

HAVE-MORE-PLAN BARN

The housing should include provision for feeding concentrates and bulk feed and drinking water. Buckets for concentrates and water should be held so that they cannot be kicked over or trodden in. Bulk feeds are best fed in wall mounted feed racks - made out of rope netting, wood slats or wide metal wire-mesh - all of which allow the goat to pull down hay etc. in a way that comes naturally to a tree browsing animal.

When designing your feed racks, there is one detail you should allow for. Goats are fussy - they ignore even the

31

tastiest food once it has fallen on the floor, yet having no hands to manipulate hay and the like, they inevitably drop a sizeable proportion of what they pull down to eat. The answer is to arrange for any food that drops from the rack to fall where it cannot be fouled, and where it can be collected later to be put back up in the racks.

OTHER EQUIPMENT YOU WILL NEED

Goats are rather low animals, so we recommend some sort of milking platform. It should be at easy sitting height and just big enough for the goat to stand on comfortably, but not big enough to let her move around and make milking difficult. There should be some arrangement for holding or tying her head at one end - a moveable vertical bar holding her at the neck works well. The alternative is to feed her the concentrates while she is being milked - that will keep her occupied. Apart from that, all you really need is a milking bucket, but it must be stainless steel. 'Optional extras' include a leather goat collar, a wall mounted bucket holder, a hoof paring knife, a coat for when she is sick, and Swiss type goat bells, all of which can be purchased from Fred Ritson, The Goat Appliance Works, Longtown, Carlisle CA6. For the modern automated goat, you can go in for a milking machine with special goat-sized teat-cups which will cost you about two hundred pounds complete, but that is for later. This equipment is made by Gascoignes and is available through agricultural suppliers.

5 feeding the goat

We cannot hope to do much more than introduce you to the
basic techniques of feeding in this book, but we want to
emphasise that, economically, this is the part of backyard
dairying that matters most. Last winter, there were some
goat keepers lashing out more than one pound per bale of
hay and ending up paying well over the subsidised $5\frac{1}{2}$p per
pint for their milk. Others were carefully tuned in to available local 'wastes' and were producing very cheap milk.

Unless you have got enough land for winter and summer
bulk food, backyard dairying is essentially a waste-disposal
industry. Putting it another way, if you have too little land
yourself to feed the goat, find an undervalued product of
someone else's land to supplement it. You are converting
things like barley straw, which you see farmers burning
every summer, into rich nutritious milk. The goat is your
convertion apparatus. What wastes should you start looking for? Use your Yellow Pages to probe local sources of:
- Pea haulms and pods (cannery waste which can be
  dried or made into silage).
- Barley straw.
- Bean straw, oat chaff etc. (less palatable than
  barley straw, but usable).

33

- Cornflour residue.
- Brewers' or distillers' grains.
- Sugar-beet pulp.
- Fruit canning and jam making wastes.

The basic principle of feeding is that a goat uses food for two purposes. Firstly, for the maintenance of her own body and necessary functions and secondly, for the production of milk. The quantities of nutrients required for maintenance depend primarily upon the weight of the goat, though they are, of course, affected by the temperature in which she lives and the amount of walking she does. The quantities required for milk production depend essentially upon the quantity of milk being produced. Now, in common with other animals, the goat needs a combination of various nutrients for both purposes: starch, protein, vitamins, minerals etc., and it would take a computer-controlled feed rationing machine to get them all dead right every day. For simplicity, farmers usually concentrate their efforts on the two most important: starch and protein.

34

THE COMMON COMFREY.

According to Mackensie, a goat needs for maintenance 0.9lb of starch equivalent and 0.09lb of digestible protein per day per 100lb of bodyweight. For milk production, she needs, in addition, 3.25lb of starch equivalent and 0.5lb of digestible protein per day per gallon of milk being given. The starch equivalent and digestible protein contents of various foodstuffs can be found by looking up in tables. Mackensie's book provides an excellent one for goats, but the N.A.A.S. (the National Agricultural Advisory Service) of the Ministry of Agriculture also produces a set in leaflet form (See Reference 7). First you work out how much starch equivalent and digestible protein your goat needs per day and you then estimate what she can probably get from grazing or browsing. On pasture, you should reckon she consumes 20lb of grass and 10lb of mixed tree leaves, gorse etc.. On rough land, reckon on her gathering 5-10lb of heather tips, gorse or whatever per day. Having then done that calculation, you can work out what starch equivalent and digestible protein you will need to supplement. See which foods give you the required total for the least outlay. Find the cheapest local bulk food available first and then supplement it with a concentrate to make up for low starch or protein. You can try 'challenge feeding'. That means gradually increasing the concentrate ration until the milk yield stops increasing. In practice, if you stay a week at a time at each level, you can judge pretty well after a few weeks what the right level is.

Apart from protein and starch, your goat will need minerals and water. Vitamins are not really a problem in goat husbandry. Minerals are usually provided by giving the goat a mineral lick in its shed. These are available from any local feed merchant. The goat takes a lick when it feels a need. Most mineral problems in goats are problems of imbalance or excess. Calcium and phosphorus are the two main mineral needs and should be balanced. Some goatkeepers favour providing licks or boxes of powder for each of the essential minerals separately, on the theory that the goat balances her own diet that way.

What we have said so far seems to suggest keeping the food rations constant throughout the year. That is not the best plan. Nor can you hope, by subtle feed-boosting, to get your goat to defy nature and produce a steady volume of milk throughout the year. One might expect a goat's feed needs to fluctuate in tune with the seasonal availability of foodstuffs. Well, they do and they don't. The trouble is that most births are in February or March, so that in fact the goat in kid needs steadily more and more to eat from November onwards. After April, her appetite gradually declines again.

Textbooks on authorship say that it is very bad practice to refer your readers to another book on your subject because what they are reading should be complete in itself. The fact is however that as a text-book on the <u>details</u> of goat-keeping, Mackensie's 'Goat Husbandry' is on its own. It would be presumptuous of us to attempt to better it. Anyway that is not the purpose of this book. We have no hesitation in advising every would-be goatkeeper to get himself a copy of this excellent work.

POISONOUS PLANTS

You should be aware that the following plants are poisonous to goats and cows:
- Laburnum
- Rhododendron
- Yew
- Beet and Mangold leaves

For antidotes, consult Mackensie or your local vet. A more comprehensive list of poisonous plants appears in a pamphlet 'Wild Food for Goats' recently published by the British Goat Society, at five pence. Many plants affect the taste of milk, butter and cheese. If you suspect that your milk is tainted the Ministry of Agriculture Bulletin No. 161 might be of help to you. In it there is a chart which relates various plants to the effect they have on milk butter and cheese.

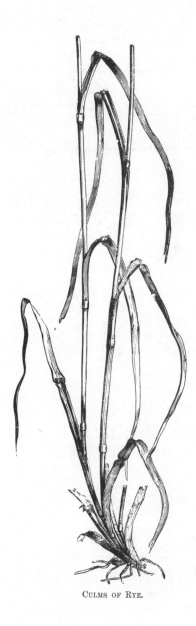

CULMS OF RYE.

37

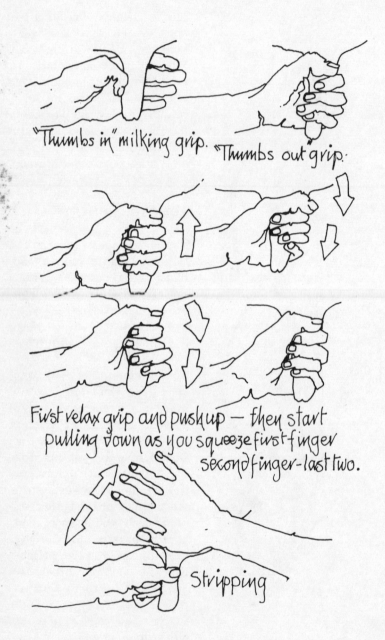

"Thumbs in" milking grip. "Thumbs out grip."

First relax grip and push up — then start pulling down as you squeeze first finger second finger - last two.

Stripping

6 how to milk

Milking is fairly simple to master on most cows or goats, although there are individual animals that are very difficult milkers. The principle is to block off the teat, full of milk, from the udder by squeezing with the index finger and thumb, and then to follow this with a squeezing downwards and out of the milk trapped in the teat, using the other three fingers. Unless you cut off the milk in the teat from getting back into the udder, no amount of squeezing will get much milk to flow. After practice, the action simplifies into an even squeezing downwards from index to little finger. The trouble is that this feels wrong at first. The natural tendency is to want to start squeezing with the little finger. Once you get over this, you should be alright. If not, you may have a difficult animal and therefore you should seek advice.

Here are some more tips:
- Let the first squirt from each teat go outside the bucket. This avoids using old milk too long in the teat.
- Massage the udder thoroughly to persuade all the milk down.
- Start using both hands. Do not be tempted to learn one hand at a time.

39

- Do not squeeze above the actual teat. This can damage the tissues of the lower udder.
- Clean a cow's udder before milking, using a swab with warm water. A goat's udder does not usually need cleaning before milking.

Although goats and cows produce best if milked twice a day, as near to every twelve hours as possible, the backyarder can make do with less production, if he so wishes, and milk just once a day for convenience. A half-way house to this approach is to use a cow partly as a milker and partly to suckle and rear calves for eating. You take the morning milk, having kept the calf out of reach of the udder overnight, and let the calf take the evening milk.

7 stage two: progressing to a cow

This section should be read in conjunction with what has already been said with reference to goats. We are assuming that you are already a goatkeeper and are about to move up to Stage Two and obtain a house cow.

WHAT COW TO BUY?

We would not recommend the inexperienced to get his first cow from a cattle market. There are no cow equivalents of the goat societies listed, although you may well, as a goat-keeper, have made contact through your society, with house cow owners. The other good source of help is the local agricultural vet who you will also have met as a goatkeeper. He will know a lot about the local dairy herds and, if you are lucky, could put you in touch with someone with a suitable cow to sell. An idea worth considering is to start with a fairly elderly cow from a commercial herd. She is perhaps no longer giving enough milk for the herdsman to keep her on, but she could be ideal for you. Scout around and see what you can find. If the first vet is unhelpful, try another. The vet will help you, too, in making sure that the cow or calf is TT-tested, vaccinated against brucellosis and free of mastitis. Do not be put off by all this talk of diseases. They are all fairly well under control now in the UK.

In our experience, there are only three breeds worth considering for a milking house cow.

     * THE JERSEY is the beautiful doe-eyed cow of cows. It is small, manageable and docile and it produces the creamiest milk of all. It is as irresistable a pet as a Siamese cat. Its disadvantage is that her bull-calves will not produce saleable meat. There is nothing wrong with the meat but there is not much of it and the fat is yellow. The housewife will not buy yellow-fat meat. However, if you are happy to consume all surplus bull-calves yourself, why worry?

     * THE GUERNSEY is very similar but slightly bigger and it gives better yields although the milk is less creamy. However the cream content is still above the average. It is said to be less temperamental than Jerseys, although we have yet to hear of temperament trouble in a Jersey.

     * THE DEXTER is the mini-cow which was the traditional house cow of the Irish peasant. It has short legs, is black and rather unattractive. On average it stands 39" to the shoulder. It is a very practical little house cow because it will eat just about anything that a goat will and manage, of course, on less than a full-sized cow. The meat is first class and produces small joints much in demand of housewives.

It all depends upon your circumstances and requirements. Jerseys and Guernseys thrive on good quality pasture, hay and silage. They need their own land. Dexters are more adaptable and can be used, like goats, partly as converters of 'wastes' from somebody else's land, although they should have acces to fresh grass for at least some of the year. They also give less daunting volumes of milk, particularly in the summer. The trouble is that there are very few Dexters still around but if you should decide upon one then contact The Dexter Cattle Society at Lomand, Seckington Lane, Newton Regis, Tamworth, Staffs. You should not have too much difficulty tracing either Jerseys or Guernseys locally.

HOUSING AND MANAGEMENT

The three basic systems apply, although rough moorland grazing would only really suit the Dexter.

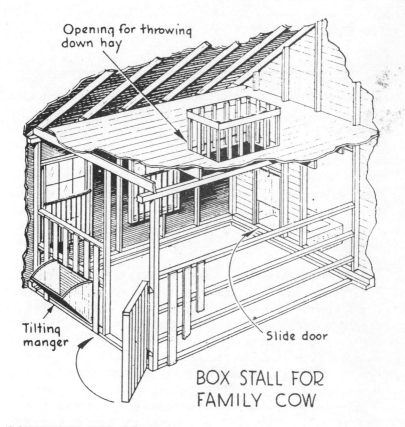

Opening for throwing down hay

Tilting manger

Slide door

BOX STALL FOR FAMILY COW

CONFINING THE COW

Cows are less likely than goats to be attracted out of a field by tasty bushes and trees, but, if they are lonely or on heat and can hear other cows, they can be almost as energetic in their attempts to get out. Give a single house cow plenty of your company and affection and keep her well confined during heat. Then ordinary farm fences and hedges should be enough, although electric fencing allows you great flexibility in rotating your available grazing. Cows need less drastic training onto electric fences than goats. Put some tasty food on the other side of the fence whilst you watch. A few jolts and she will get the message.

HOUSING THE COW

Again the needs are similar - just larger. The great advantage, however, is that a cow can happily be left out summer and winter, day and night, if you so wish. Milking can also be done outside, but it is as well to do it under cover in the wetter weather.

FEEDING

The same principles apply as before, although Jerseys and Guernseys are much more fussy about 'wastes'. For these breeds, hay and silage are the only really acceptable bulk feeds.

Unfortunately there is no bovine equivalent to Mackensie's 'Goat Husbandry'. The book that comes nearest in our opinion is Newman Turner's 'Herdsmanship', also published in the UK by Faber and Faber. Although this book is designed for the commercial dairy herd owner, its approach is organic and there is plenty of useful information for the one cow beginner. Unfortunately this book is now out of print but your public library should be able to locate a copy for you to loan. Another helpful book to as your library for is Kenneth Russell's 'Principles of Dairy Farming'. The Ministry of Agriculture also do a useful bulletin called 'Rations for Livestock'.

SHARING THE LOAD

We have already emphasised what a tie it is having either a goat or a cow. Having a cow, you are also faced with having to make use of pretty large quantities of milk. Unless you

44

have got all the attributes of an officially approved dairy, you are not permitted to sell the milk. But what is wrong with owning a cow in partnership with friends? A good scheme we know of is run by two couples, one with two acres and the other in the nearby town. They share the milk and the feed bill half and half but, for their work, the country people get the calf every year as well. The only drawback to this idea is that you must have some method of getting the milk to your partners cheaply at least every other day.

8 dairy products

Milk is a pretty good food as it is. In fact, there are some primitive peoples who live on virtually nothing else. So why bother converting it into other things? The traditional reason must have been to produce from it foods that could be kept for the winter, but nowadays this reason hardly applies since adequate supplies of milk are commercially available all the year round. No, butter and cheese are still made today because they represent concentrated sources of fats and proteins which it is economic to transport over longer distances than milk and because they have their own consistencies and flavours prized independently of milk.

The backyard dairyman can, if he so wishes, obtain excellent year-round food from milk alone (assuming he has plenty of winter feed for his animal) but butter and cheese will further enrich his life. Not only that, but instead of throwing the inevitable surplus summer milk to the pigs or chickens, it can be converted into concentrated foods which can be stored.

But dairy work is time consuming. It has to be done with reasonably fresh milk, of which the home dairymen will have limited quantities, and the trouble is that it takes almost as long to make butter and cheese from five gallons of milk as from fifty. So is it worth it? At today's subsidised prices, you get a low return on your labour time for making your own. But they will be superior, after practice, and greatly satisfying. Cheese making is as absorbing a hobby as home wine making and, in many ways, very similar.

46

47

9 a little technical information

HOW IS MILK PRODUCED ?

Milk is produced by all mammals as a food for their young, during the early part of their lives when their digestive systems are not able to take other foods. It is manufactured in the mammary gland of the mother animal. Production is induced just before the mother gives birth - ready to give the young their important first food. Actually, this first milk is rather special. It is called 'beistyn' - it is golden yellow and as thick as double cream. It is designed to clear the inside tracts of the newborn animal. In England, there is an old law that beistyn must not be sold, but often a cow or goat will produce more than the infant animal needs. Even diluted one to four with ordinary milk, it will set like an egg custard. The taste of beistyn is an added bonus for the home dairy producer.

Production of milk continues regularly in the mother animal for a period known as the lactation period. This varies considerably in length, but the point is that if you want your animal to continue giving you milk, you must have it mated regularly to make sure that it stays 'in milk' for a good proportion of its life. Although both cows and goats have been

48

known to continue a single lactation for three years and longer, most people reckon it is safer to get the animal pregnant regularly every year.

An animal will only stay in milk so long as all the milk she produces is being used. This is nature's regulating system. In the normal course of events, the mother would thus only continue giving milk until its young start eating other foods. If you milk the animal dry twice every day, she will continue to produce a good quantity for longer, but you must keep milking dry - known as stripping.

The mother can only really be milked if the udder is 'let down'. Some outside stimulus, such as the cry of its young or the arrival of its owner, induces the production of a hormone, oxytocin, in the pituitary. It takes about one minute in a cow for the oxytocin to reach the udder and cause the muscles to relax to let down the milk. The hormone continues to be sent down for about a further eight minutes, during which time you should have completed the milking. The important thing is not to stop for breaks. Once you start, you have to carry on until the animal is dry.

The process of milk production in the udder is not yet fully understood, but it appears to be a combination of filtration of certain constituents from the blood-stream, cell degeneration and most important of all, synthesis of other constituents by cell metabolism.

The simplest mammary gland known is that of the Australian duckbill platypus, where droplets of milk ooze out of a small hole onto the animal's hair, where they are licked off by the baby platypus. The udders of the goat and cow are far more developed, with milk tracts leading down from the main producing tissues in the gland, contained in the milk sac, to a reservoir just inside from the teat. The teat is a sort of automatic portion dispenser, like those used in bars for spirits, making sure that the infant only gets down as much milk as he can swallow comfortably. The udder of the cow is divided by thin membranes into four quarters, each with its own teat and each needing to be milked separately. The udder of the goat is divided into two sections.

WHAT IS MILK?

'Milk is an oil-water emulsion, the continuous phase being aqueous.' (see Reference 10) That means that it consists of water with certain substances dissolved in it, plus little fat globules floating around in it. The proportions of the various constituents in milk vary considerably, but an average make-up of scientifically analysable materials is:

> 87% water.
> 4% fat.
> 5% milk-sugar or lactose.
> 3% caesin.
> $\frac{1}{2}$% albumin.

(These last two are the proteins in milk.)

In addition, there are bacteria, many of which arrive in the milk during and after milking, some of which are useful in the human stomach or in the process of cheese making, but others of which are harmful. There are probably also loads of trace elements in milk, necessary for good nutrition but not easily analysed out. Milk is known to be rich in calcium and in vitamins A and B. It also contains phosphorus, potassium , iron and vitamins C and D.

WHAT IS CREAM ?

The little fat globules rise to the surface of settled milk and form cream. Eventually, the fat globules will settle at a certain concentration at the top of the milk and the familiar cream line appears. This process can be speeded up in a centrifugal separator. The butterfat content of cream is very variable - between 10 and 65% fat, but 45-55% fat is a more normal level.

50

WHAT IS BUTTER ?

Butter is formed by persuading virtually all the fat globules
in the milk to join up into a continuous mass, separated from
the water and its dissolved constituents in the milk. The
persuasion that seems to work best is severe agitation at
controlled temperatures, but the exact mechanism by which
it occurs is still a matter of dispute. Butter is made up,
very roughly, of:

      84% butterfat.
       7% water.
       7% unremoved buttermilk.
       2% salt, added for flavour.

INTERIOR OF A DUTCH DAIRY.

51

The 'buttermilk', left behind after the butter separates out, is made up of:

> 90% water.
> 5% milk-sugar.
> 4% milk-proteins.
> 1% butterfat.

WHAT IS CHEESE ?

If the acidity of milk is increased, the separate particles, particularly of milk protein, coagulate together into a jelly-like or custard-like substance called curd. A great deal of the water and a little of the dissolved substances separate out into a greenish watery liquid called whey. This separation takes place more easily at above sixty degrees Fahrenheit. The increase in acidity can be achieved by one of three methods.

     1. Simply add an edible acid, such as lemon juice or vinegar, to warm milk.

     2. Let warm milk stand where it stays warm. Little organisms in the milk (Streptococcus Lacticus) convert the milk sugar, lactose, into lactic acid and hey presto, the milk curdles by itself. But if you leave it too long, another load of organisms take over and turn it into 'bad' milk, so take care.

     3. The conversion of lactose to lactic acid can also be achieved by using enzymes. Rennet is the usual one employed, It is taken from the stomachs of calves and processed commercially. Very small amounts of commercial rennet added to warm milk will produce curd.

To make cheese, the curd is separated from the whey. It is then eaten fresh as soft cheese, or pressed into moulds to dry out the whey and water completely thereby leaving it in a form which will keep for longer.

Cheeses vary enormously in composition but here is a rough

guide:

%	Camembert	Cheddar
Water	45-51	27-34
Fat	21-30	26-30
Proteins	18-23	27-40
Lactic Acid	0.5	1.5

(from Richmond's Dairy Chemistry)

Cheese is valuable, not merely as a way of storing milk for winter, but in its own right as a concentrated protein food. The other great thing about cheese is that you can make so many delicious variations.

There are plenty of variables in cheese making:
- the method of increasing the acidity.
- the temperature at which the curd is formed.
- the acidity at which it is formed.
- whether is heat is used to produce further separation.
- how much heat and for how long.
- whether the cheese is pressed.
- under what pressure and for how long.
- whether the cheese is affected by other bacteria.

We will have more to say about the use of these variables in re-creating famous traditional cheeses later in the book.

HOW SOME OTHER MILK PRODUCTS ARE MADE

HOMOGENISED MILK is milk, heated to reduce the surface
tension of the fat globules, then forced through very small
holes at high pressure to reduce the globules to a size at
which they no longer rise.
CONDENSED MILK is milk, mixed with sugar, then heated
to around 235 degrees Fahrenheit and condensed in a vacuum.
DRIED MILK is concentrated milk. This is achieved by blow-
ing hot air through the milk, thereby concentrating it. This
concentrate is then placed on a rotating drum which pushes
out the rest of the water. It becomes brittle solid and is
then ground into powder. It is now also produced by a freeze
drying process. Some dried milk is dried whole milk, some
is dried skim milk.
TINNED CREAM contains, on average, half the butterfat
content of fresh cream. It is sterilised at 250 degreesF
and should keep for a year in a tin.
CASEIN is sometimes separated out in large dairies. It is
used to make nice white buttons and nowadays for that re-
volting artificial milk powder offered in sachets by mass-
production caterers as a substitute for milk in coffee.

10 cream production at home

Cow's milk separates into cream and skimmed milk fairly
easily under the action of gravity alone. Goat's milk does
not and the only effective way of producing fresh cream from
goat's milk is by means of a separator. Frankly, this is a
bore. But if you are a goatkeeper and sufficiently interested
in cream, then try to pick up an old farm hand-separator by
asking around at local farms or by advertising. The cheapest
new separator still made is a seven gallon model at over
three hundred pounds - out of the question.

Before you buy an old one, make sure that the tinning is unbroken and that all the cones are there still. The separator
gives 99% of the available cream and you can adjust it for
the thickness of cream you want. The trouble is that even
the simplest machine has about twenty three different parts,
all metal, which need to be washed in soda and sterilised
after each use. Even if you use a dairy disinfectant rather
than boil the components, it is quite a chore when the amount
of cream involved is small.

If you are the happy owner of a cow, then cream making is
simple. There are two methods for home production.

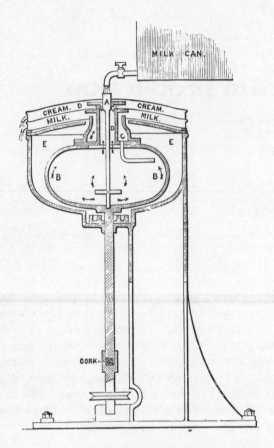

-Section of Separator.

THE SETTLING PAN METHOD. You need a big wide shallow pan (not plastic), and a cream scoop (a saucer will do), to draw off the cream once it has settled. Leave it for twenty four hours in a cool place (longer for goat's milk), then skim off the cream, starting by breaking it off at the edges of the pan first. It comes off like a thick skin. This method gives you around 90% of the potential cream available. It needs very little equipment, but quite a bit of cool shelf space.

56

THE SYPHON METHOD. (This was developed by Len Street). For this you need gallon jars, preferably caterers' salad cream jars because they have wide necks, and a wine syphon from Boots or other chemists. Slice an oblique cut at the end of the plastic pipe on the syphon. Keep the milk in your gallon jars for twenty four hours in a cool place. Then slide the syphon pipe down inside the edge of the jar past the cream layer to the bottom of the skim milk below. Syphon off the skim milk , finishing with the jar tilted over until you reach the visible cream line. You are left with cream in your jar. This method involves less storage space and the equipment is easier to get than a settling pan , but it is a bit more fiddly.

Of these methods we favour the syphon. The jars can be placed straight in the fridge: This speeds settling and gives a better-tasting cream than the settling pan method, because the large surface area of the milk in a pan tends to pick up taints.

If you have a fridge, there is no need to make cream every day. The milk will store for three or four days, so you can get away with making cream twice a week if you wish. How much cream you get varies, but it takes about eight to ten pints of milk to give one pint of cream.

The skim milk can be drunk fresh, but if you have enough whole milk left over for drinking, you won't want to bother with the thinner stuff. It is an excellent food for feeding to calves, kids, chickens, pigs and other animals. Other than that, it is fine for turning into yogourt or skim milk cheese (see recipes later on).

CLOTTED CREAM

Clotted cream makes a pleasant change for 'cream teas' and the like. It keeps longer then fresh cream and so is better for butter making. It means you can spread your butter-

making sessions out to once a week or more. You may find
that clotted cream is the only foundation that works for butter-making when the milk starts to get thin in the autumn.
If you keep a goat, clotted cream is the only type you can
make without a separator.

Stand the milk in a heatproof pan to a depth of six to eight
inches, in a cool place for 24 hours. Transfer the pan gin-
gerly to your stove and heat up gently to 150-170F for use
as clotted cream. It will form a ring at the edge of the pan,
and the cream will split away. Let the pan stand to cool for
12 hours, then skim off the cream. It is not a lot of work,
but it does take a long time.

11 butter making at home

WORKING THE BUTTER.

Here is a very simple butter-making method from Jim Platts of Willingham, Cambs:

> "We don't own a milking animal. We buy ordinary milk (not Gold Top) and take the cream off each bottle using a 5ml. plastic syringe. (Your doctor will throw away several of these every day). We put the cream in a two pound Kilner jar (the wide top makes it easy to get the butter out), and when it is half full, we stand it near the fire until it is just warm to the touch. Then we shake it by hand

59

until pellets of butter form. We reckon usually on less than five minutes' shaking, and a yield of $\frac{1}{2}$oz. of butter per pint of milk. We produce as much butter as we want from our weekly milk supply."

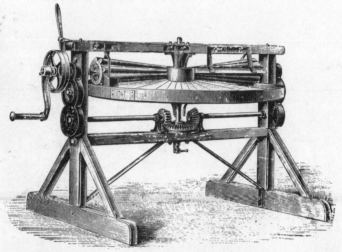

COMPOUND BUTTER-WORKER.

Now, for those that want it, here is a little more detail, and some ideas on scaling up the process to use kitchen machinery. There are three types of cream that can be used to make butter.

FRESH CREAM is fine if kept refrigerated prior to butter-making. Alternatively, it can be kept by adding at least 8% by weight of salt - but this produces a very salty butter not to everyone's liking.

CLOTTED CREAM keeps well before the next butter-making session and the taste of the butter obtained is particularly liked once people get used to it.

SOURED CREAM (or Ripened Cream) makes a butter with a distinctive soured flavour. Making butter with soured cream is easier - it turns granular more quickly. For a start, we suggest you have a go with fresh cream. If you experience difficulties, or feel like a bit of variety, try one of the other two.

Keep the cream good and cool before butter-making. There is nothing to beat a 'fridge. Leave at least twenty four hours between the last addition of cream and the butter-making session. Stir the cream jar well each time you add a new load of cream.

You do not need to use a special butter churn for butter-making, but they are good for larger quantities. If you do not want to track down a churn, and if you are lucky enough to have an electric food mixer, then you have got nothing to worry about. But first get your cream to the right churning temperature:

52-60F in hot summer weather
58-66F in cooler weather

For this, we suggest you invest in a proper dairy thermometer.

Take a cupful of the thickest cream first and spoon it into the mixer, working at a slow speed (around 90 r.p.m.), with a cake-mixer head fixed. The bowl and head should be cold. Watch until patterns are made, rather like those that appear with whipped cream, then add another cupful of the cream and repeat until all the cream is in. Carry on mixing until the colour changes to golden yellow, and then until the wheat-sized granules of butter are formed.

BUTTER-WORKER.

You can then just squeeze the butter-milk out in a muslin bag. This will remove most of it and give you butter fine for the next week's use, but if you want something you can store, refrigerated, for months, you had better wash it and work it well.

To wash the butter, you will need water at about the same temperature as the buttermilk. Pour some into the granular butter in your bowl and carefully fold the granules over to wash out all the buttermilk. Strain off the wash-water and repeat until it stays pretty clear. You will probably need about five washings to get to this stage.

After the final straining out, sprinkle on some salt if you like - we suggest one dessertspoonful per pound of butter. For the really professional touch, add some annatto butter colour - particularly if the butter looks less golden-yellow than the shop-bought stuff you are used to. It probably will do, except in the very best grazing periods of June and July. (See the cheese section for how to make your own colouring - and for addresses of annatto suppliers).

If you want the butter to keep, its water content should be

ECCENTRIC CHURN.

62

NOVEL METHOD OF CHURNING.

under 16%. Commercial butter is more likely to be around 12%. To remove the water, work the butter. The simplest method to start with is to use a cutting-board and a strong wide-edged knife or spatula. Squeeze down the butter onto the board, forming a thin layer. If you slope the board (say, on the edge of the draining board), the water you squeeze out will run off immediately. Fold it over and repeat the squeezing down, without rubbing or smearing the butter. Keep it fairly cool during working and keep going until you really are not getting much more out. Later on, you may feel like buying or constructing something more elaborate in the way of a butterworker, but that is beyond the scope of this book. Whatever you use, keep it spotlessly clean. It is best to keep a special board for working, rather than the one you used yesterday for crushing the garlic.

BUTTER MAKING PROBLEMS

There is really only one big problem that occurs in butter
making and that is for one of many reasons, the butter just
does not 'come' in 20-30 minutes. If 40 minutes have passed
and you have still got white cream, it could be for one of the
following reasons:

* Your cream is too warm or too cold - sometimes
  in winter it really needs to go up to 65-70 F.
* Your equipment may not be really clean and sterile.
  This may cause what is known as 'ropy fermentation'.
* The animal may be almost through her lactation -
  in which case, butter is harder to make.

Here are a couple of other tips we have heard but not fully
tested out. If goat's cream gives you trouble getting too big
a granule size, add a cupful of water per one and a quarter
pints of cream as soon as the granules appear. Also, it is
said that thinner cream is better in summer and thicker
cream is better in winter.

Do not throw away the buttermilk. It contains the milk-sugar
and milk-proteins. Drink it fresh, use it for bread making
instead of water, or make cultured buttermilk, a fine-tasting
product which you pay heavily for in the shops. Add one part
of ready- cultured buttermilk, either from a previous load
or shop-bought, to eight parts of new buttermilk. Let it
thicken in a warm place and then store it in the fridge. Alternatively, you can use a commercial starter or lactic acid
starter, which has the advantage of keeping longer.

12 cheese production at home

'MILK IS IMMORTALISED IN CHEESE'

(Enoch Powell)

As children of the super-industrial state we have been cond-
itioned to the standardised artificially cultured cheese of
uniform flavour and consistency. But just consider, before
you start home production, how things were before the Industrial Revolution. Each cheese depended upon the quality
of pasture, both in terms of its geography and its season; it
depended upon the size of the farm and whether or not the
cheese was made for sale (in which case, even then, it had
to be more uniform). This infinite variety was at the basis
of the many 'standard' varieties of cheese produced now in
rigidly controlled conditions to ape the way the cheese came
out naturally from certain farms.

For instance, 'moorland' cheese was only made in Spring,
when the cows emerged from winter quarters and grazed on
the young scented herbs and grasses of the moors. This diet
cleared their blood and they gave a delicious casein-rich
milk which in turn produced a beautiful, light textured, blue
cheese. The moorland people would never have dreamed of

65

making this cheese at any other time of the year - but a
cheese factory has to keep going throughout the year.

Take another case. The blue-veined Stilton depends for this
effect on the action of the mould Penicillium Roqueforti. The
cheese rooms in the Stilton district of Leicestershire were
well infected with this mould and the process carried on
naturally. Nowadays, the milk for Stilton making is heated
to destroy the existing organisms in the milk. Penicillium
Roqueforti is then injected in measured quantities to ape the
conditions of these old Leicestershire cheese rooms.

The point is that the characteristics of each local type of
cheese depended originally upon the peculiarities of the local
conditions to a very large degree. We could, in this book,
give you a great stream of recipes telling you how to ape the
local cheeses of evry district of England, France and the
rest of the World. Instead we have chosen to introduce the
basic processes common to most cheese making, plus some
variations developed in certain areas and worth a try in your
dairy. We hope this will spur you on to produce unique individual cheese which varies over the year and with your experiments. Maybe, after a while, you will decide to stick to
a particular variant which pleases you best.

We have already been through the basic chemistry of turning
milk into cheese. Now we will proceed to some practical
recipes for making cheeses. You will remember that there
were three ways of creating the necessary acidity for curd
formation. Here are basic recipes for each way:

ADDING ACID. This is the method used for cheese
making in India. It produces a soft, rubbery cheese called
'Panir' which is chopped into 2" x 2" x $\frac{1}{4}$" blocks, fried or
added to curries. 'Panir' is also used to make certain Indian
sweets.
Heat one and a half pints of milk, stirring, to boiling point.
Remove it from the heat and add quarter of a teaspoonful of
tartaric acid ('Cream of Tartar') dissolved in a cup of hot

water. Stir until the whole of the milk curdles. Leave it for fifteen minutes covered and then strain it through muslin and squeeze out all the whey. At this stage it is called 'Chenna'. To make 'Panir', keep it in the muslin, tied tight and place it under a 5lb weight for two hours. For more details on how to use this in Indian cookery, get hold of Mrs. Balbir Singh's 'Indian Cookery', published by Mills and Boon.

LETTING THE MILK SOUR. This is the method behind all the various so-called 'lactic' cheeses (meaning cheeses soured by lactic acid rather than rennet). It is a simple method for the home production of very pleasant soft cream cheese. This can be made from ordinary soured whole milk or from soured buttermilk, or from yogurt. Any of these substances, as long as they are thick, can be drained in a muslin or cheesecloth bag to produce a form of soft cheese. It helps to open the bag after half an hour and remix the contents. Repeat this again after another hour. By doing this, a more even, creamy cheese is produced. After that, there are further possible variations. You can press the cheese between plates for an hour or so, or add some fresh cream until the cheese works up into a fine pat of cream cheese. Salt the cheese to your taste and maybe end up by adding some form of extra flavour. The possibilities here are endless. Try starting with chives, wine (stirred in), orange slices or chopped nuts.

The cheese made by draining yogurt overnight is very popular in Germany under the name of Quark, and is quite delicious used as we would use whipped cream on fruit.

ADDING RENNET. This is the method for all hard and semi-soft cheeses and for a lot of the better soft cheeses. This is where cheese making gets interesting but rather complex. However let's start with a very popular French cream cheese called Gervais.

For every gallon of milk, you will need a quart of cream.

Mix them and heat to 60-65 degrees Fahrenheit while stirring. Add ½cc of commercial rennet diluted 1:6 with water. Stir for 3-4 minutes and then leave for ten minutes. Ripple the surface with your fingers for about one minute. This prevents a skin of cream forming but allows curdling to proceed. Ripple again after thirty and sixty minutes. Then leave it overnight for twelve hours at 60-65 degrees Fahrenheit. In the morning it should be nicely set. Ladle it into a cheesecloth bag and let it drip for thirty minutes. Mix the contents and re-hang it. Mix it every hour for six to twelve hours until it has the firmness of Gervais. Salt it to your taste and finger-press it to the shapes you want. Store them in foil in your fridge. It is best to eat after a week.

This recipe is based on a master recipe for cream cheese in John Ehle's 'Cheeses and Wines of England and France', a unique book on home cheese making with excellent recipes for all the well known types. It is published in the USA by Harper and Row. Let's hope that it will be available soon over here.

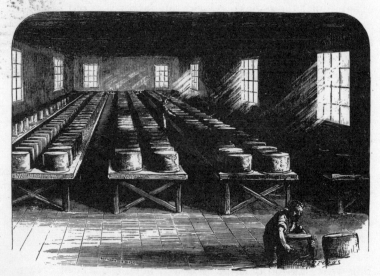

CURING-HOUSE, WHITESBORO' CHEESE FACTORY.

This basic recipe can be varied. It can be made without cream or with more rennet and thus less setting time. Ring the changes as you wish.

HARD CHEESE

Although there are a great many variations, the making of hard cheeses generally follow this basic ten step pattern.

1. RIPENING THE MILK. This is a delicate process and one where you will need to experiment according to your own conditions and tastes. If you let the milk sour too much, you may end up with a curd which is an acid curd/rennet curd mixture, difficult to ripen. If you use milk too fresh, the acidity is inadequate for the rennet to have the desired effect.

The traditional farmhouse solution is to let the evening milk ripen overnight at a temperature of 50-60 degrees F. In the morning, warm it and stir the cream back into the milk. Add the fresh morning milk and your cheese making milk should be about right. You can, alternatively, add commercial starter (lactic acid culture) to fresh milk and let it stand in the warm to ripen it. Whatever you do, the milk should still taste sweet.

2. COLOURING THE MILK. This is optional but red annatto is added to certain cheeses. This is up to you. You could experiment with using saffron or marigold petals or the juice from pressed grated carrots. This is what used to happen before annatto was imported.

3. COAGULATING THE MILK WITH RENNET. Extract of rennet is stirred in at around 85-90 degrees F for most cheeses. Some cheeses are rennetted as high as 105 degrees Fahrenheit. The usual practice is to add the required amount of rennet, diluted 1:6 with warm water. So long as the acidity is right, this will produce a curd hard enough to hold the fat globules together without being tough. Rennet curd does shrink naturally and this helps to expel the whey. The higher

69

the temperature and the acidity, the faster this shrinking takes place. If there is far too much acid in the milk, it will eventually interfere with the shrinkage process and you end up with a nasty wet acid mess. Make sure that you stir the rennet well in and then leave the milk undisturbed at 85-90 degrees F until it has set to the consistency of junket. This may take anything from fifteen to sixty minutes. The usual way to test curd is to poke your index finger under the surface, then lift. If the surface breaks cleanly, then it is ready.

Nowadays, you can buy 'acidmeters', which are miniature testing kits, easy to use, for testing the acidity of curd. Alternatively, John Ehle describes an old method based on sticking pieces of curd onto a hot iron. (See Reference 12)

4. BREAKING THE CURD. You can leave the curd unbroken to drain, but the process is speeded up if you break the curd into pieces. You get better results if you keep the pieces of curd even in size. Most English varieties of cheese are made with quarter to half inch cubes. The illustration shows one way of getting fairly even cutting with an ordinary kitchen knife.

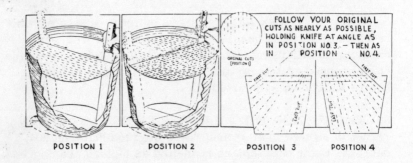

POSITION 1   POSITION 2   POSITION 3   POSITION 4

The difficulty of producing cubes of even size right the way down can otherwise be achieved using a curd knife which is obtainable from Clares Carlton Ltd., Wells, Somerset.

5. TREATING THE CURD. The purpose of this stage is to firm the curd more effectively by heating it in the whey. It is also to improve the drainage of the whey from the curd by keeping the pieces of curd gently moving. You can carry out both treatments at the same time, or separately. If you break the curd with a knife, it is perhaps better to start by stirring the curd gently with your hand for about fifteen minutes. This way, you can check that the pieces are all fairly even and break up those that need it. Try to stir as gently as will keep the curd pieces from sticking together. After this, you can heat the pan very gently to 100 degrees F over about an hour, stirring from time to time to keep the pieces from sticking. The curd pieces should then have firmed up sufficiently to fall apart in your hand without squeezing. If you keep them at 102 degrees F for another hour, this will firm up the pieces even more so that a handful of pieces will shake apart after being pressed together in your hand. This is the degree of separation you should aim at as a beginner before you go on to drain the curds. Later you may wish to shortcut this rather long process. Do not leave the curds and whey together too long because the acidity starts to rise and the curds go wet and unmanageable.

6. DRAINING OFF THE WHEY. This is most simply done by pouring the curds through muslin. The only skill is in deciding when the curds are ready to be drained.

7. TREATING THE DRAINED CURD. This very important stage determines to a large extent the final texture and ripeness of your cheese. The degree to which you further break up the curd at this stage, by cutting or squeezing, determines its texture. If you like crumbly cheese, keep the pieces loose. If you like it more resilient, squeeze the curds together into more solid blocks. Before you do any of this, you must however salt the cheese. One tablespoon per sixteen pint load is a good starting guide. The main point of adding salt at this stage is to stop the action of the acid-producing baccillae. For certain cheeses, you chop or mill the curd at this stage as well.

71

8. FORMING THE PIECES INTO CHEESES. At home, this is most easily done by forming a 'bandage' of cheesecloth around a cake-shaped mass of curds about 6" across, and pinning the bandage in place. With a couple of thicknesses of cheesecloth above and below the 'cake', it is ready for pressing. Alternatively, you can make your own wooden cheese mould out of wooden storage jars, available from kitchen supply shops, with a few holes drilled in the sides and bottom to drain out the whey. This kind of mould, however, will not stand much pressure. You could also try making cheese moulds from old stainless steel saucepans or large food cans with holes drilled in them. Whatever you use to contain the curds, make sure that they are packed in well, with no cracks extending into the centre of the cheese. Again, the shape and size of the mould also depends upon which regional cheese you are trying to copy.

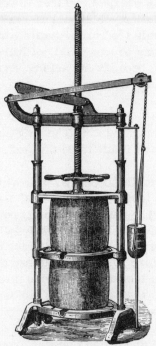

CHESHIRE CHEESE-PRESS.

9. PRESSING THE CHEESE. If you can get hold of an old cheese press, so much the better. If not, any simple device will do, providing it gives an even down pressure onto the cheese, which continues to be even when the cheese shrinks down. At its simplest, this can an empty can which slides down into a can mould, with two bricks placed on the inner can. The usual methods of pressing involve an initial pressing (you can start with two bricks), then an upending of the cheese and a second pressing at a higher pressure (try four bricks). The period of pressing can also vary, but we suggest you start with twelve hours at each pressure. This should be fine for a small six inch cheese but naturally the cheese factories these days press their products under pressures of several tons to speed up the process.

10. RIPENING THE CHEESE. The fermentation process that takes place at this stage can give the cheese a fine mellow flavour or can spoil it completely. Be warned that, for no apparent reason your cheese may dry into a pile of acid crumbs or it may stick to the shelf in a slimy mess, or it may even roll off the shelf completely - propelled by the gas of its own fermentation.

The cheese should be transferred to a cool but frost free cellar or pantry. Beforehand, you may choose to protect the cheese once the rind is dry by painting on a layer of paraffin wax heated first in a double saucepan. You can also use lard or cornflour. Ripening cheese should be turned regularly every day for the first few days, then every other day until the cheese is hard and ready - usually in three or four weeks. You can taste it at intervals, if you wish, by taking out a small piece with an apple corer. Make sure you put back the outer plug, and seal the wound again with wax, lard or cornflour. Most cheeses are best then left until at least two months old.

RINGING SOME CHANGES

* Add butter to the drained curds

* Try 'green cheese' that is cheese freshly pressed but not ripened.
* Put leaves inside the press before you fill it with curds: dock leaves, nettles, vine leaves are some which are used to wrap cheese.
* Try letting some of your cheeses go 'blue' by leaving them up to six months. Depending upon the organisms present in your pantry, interesting may start to happen.
* When salting, grate some cheese of the desired type into the curds. This sets up a good bacterial growth for the ripening stage.

If you are a vegetarian, you may want to try using casein rennet rather than the usual stuff. We found this recipe for making your own:

Set six pints of milk as for clotted cream. Skim the cream off. Add two tablespoons of vinegar to the skim milk and heat it until it curds thoroughly. Wash the curd in water three or four times, kneading thoroughly. Dry the curd and powder it. This powder is almost pure casein.

SIMPLE CHEESE-PRESS.

74

SUPPLIES

We know of only one supplier of a good range of equipment for home cheese making:

Clares Carlton Limited,
Wells, Somerset BA5.
or at:
7, Winchester Avenue,
Denny, Stirlingshire.

curd knife

They supply acidmeters plus chemicals and spares, floating dairy thermometers, cheese samplers, curd knives, shovels, scoops, whey bowls, textiles,and Coulommier cheese moulds Nobody, to our knowledge, supplies cheese presses.

For rennet and annatto, you can choose from Clares Carlton
or: R.J. Fullwood and Bland Limited,
        Ellesmere, Shropshire.
or: Chr. Hansen's Laboratory Limited,
        476 Basingstoke Road,
        Reading RG2 OQL
Hansen's also supply cheese cultures and special cheese wax.

13 yogurt production at home

Fifteen years ago, if you had done a survey in this country, I doubt whether 10% of the population would have known what yogurt was. All that has changed now. Big food manufacturers recognised its potential as a bland, creamy textured, eminently suitable for flavouring with lots of different sweetened fruit flavours and also having a 'health' image. Yogurt must have been one of the biggest food success stories since baked beans.

Needless to say, the traditional product is rather different from what we get from our supermarkets - as you will know if you have ever tried it in Greece or the Middle East. Traditional unboosted, unsweetened yogurt has been popular throughout Northern and Central Europe and Western Asia, but is supposed to have originated in Bulgaria. There are many local variants, from SKYR in Iceland, through DAHI in India, TAETTE in Scandinavia, LEBEN in Egypt, to MAZUN in Armenia.

Traditional yogurt is milk coagulated by the use of added cultures into a soft curd (see the previous section on "What is Cheese?"). In hot countries, culturing is a way of pre-

76

serving milk for slightly longer in a palatable form. In addition, it has had ascribed to it various therapeutic properties, mainly, it appears, because of the unusual health and longevity records of people in certain countries where it is eaten regularly.

The cultures normally used are LACTOBACILLUS BULGARICUS and STREPTOCOCCUS THERMOPHILUS, and the traditional idea is that regular consumption of yogurt establishes a colony of these micro-organisms in the intestines and that they assist digestion. However, some recent research suggests that these particular organisms are not, in fact, able to establish permanent colonies - they just help things along on their way.through. This means that you have to keep eating yogurt regularly for it to be beneficial.

There is more to come. New milk curdling cultures have been discovered which occur both in mother's milk and in the stools of infants. One is LACTOBACILLUS ACIDOPHILUS. This also occurs naturally in cow's milk. This culture is known to be particularly beneficial in that when established in the lower intestine, it produces conditions unfavourable to the growth of harmful putrefactive bacteria. It also aids the breakdown of food particles and the synthesis of vitamins B and K.

The trouble is that these friendly intestine dwellers are killed off when you take penicillin and most other antibiotic or sulphur drugs. So, if you do have to take these kinds of drugs, it might be an idea to follow them up with a few doses of yogurt made with Acidophilus. Certain strains of Acidophilus have now been developed which are resistant to a range of antibiotics and which can be taken in powder form or made into yogurt. The stuff to buy is ENPAC, made by Aplin and Barrett Ltd. and available at chemists. Acidophilus, though for use as a yogurt culture, will only really work on skim milk.

Whether you use Acidophilus or the more usual cultures,

the principles are the same. You add a small quantity of culture, or milk from a previous batch containing culture, to milk, and then you incubate the cultured milk until the culture has thoroughly multiplied and coagulated the milk to the desired consistency.

Any plain bought yogurt will work as a starter. Do not be fooled by the labels in healthfood stores saying 'LIVE yogurt'. All commercially sold yogurt, at least in the United Kingdom, is live.

The are many variations possible on this basic method. If you are using some shop bought yogurt as a starter, add it in at about one tablespoonful to the pint, mix it in well, and hold it for three to four hours at 70-105F. If you buy culture stick to the supplier's directions on time and temperature.

DAHI is based on buffalo's milk, which is very rich; and a lot of Middle Eastern yogurt is based on goat's milk, which makes a very sharp, rather thin yogurt - very fine. But my alltime favourite is the YAORTI you get in Greece, made from ewe's milk. It has a beautiful, light, yellow, creamy crust, and is firm and jelly-like. Served with a little clear honey on top, it is out of this world..... I don't know how they get that crust on it - whether it is only possible with ewe's milk, or whether it is a special local culture. If ever I go to Greece again, I will find out.

If you or your children are hooked on supermarket yogurt, that's not difficult to copy either. Add dried skim milk or dried whole milk (whatever it says in the 'ingredients' printed on your favourite brand) to the milk you use. You will need to experiment to get the right final consistency. For flavouring, use a good full-fruit jam and mix it up well. This way you can make a 'modern' yogurt for under half the supermarket price.

A NORMAN DAIRY.

Keeping a batch of cultured milk at around 80-90 F for a few
hours needs a bit of care. The simplest method is to buy a
big vacuum flask. Warm it first. Heat the cultured milk in
a saucepan to blood heat and put it in the flask. We leave
ours overnight, but that is only out of laziness. It is less
acid if you cool it right down once the culture has formed a
complete curd, (usually three or four hours).

If you like your yogurt jelly-like in individual pots, then em-
bed the pots in a box of polystyrene beads (sold for making
'sack' chairs) or vermiculite roof insulation chips. Cover
the filled box with something equally insulating like polystyrene sheeting, or a blanket. Do not let the polystyrene beads
get in the yogurt - they taste most peculiar, and cannot be
good for you.

We definitely do not think that the expensive electrically
heated 'automatic' yogurt makers are really worth the ex-
pense, especially when you can do the job so simply without
them.

14 yet more dairy products to make at home

SWEET PUDDINGS

JUNKET is just milk set with rennet - the basis of cheese - eaten fresh. The rennet used is in the form of 'junket tablets' and can be bought from good grocers that way. Follow the directions on the packet, but basically all you have to do is heat milk to around 90 degrees F, add the tablet and leave it to set.

The traditional English flavouring for junket was rum, but powdered nutmeg or cinnamon on top also helps. Personally, I find junket a very dull dish compared to either yogurt or blancmange.

BLANCMANGE. The addition of two and a half tablespoons of cornflour to a pint of milk, and then boiled and cooled to set turns the milk into a very acceptable pudding. It can be flavoured a hundred ways, but almond essence and chocolate are my favourites.

The home production of ice cream is not as widely accepted here as it is in the USA, partly because our summers are not as hot and partly because so few of us have yet discove- red the vast superiority of ice cream made in a special ice

80

cream maker, rather than in the ice tray of a fridge. Unless it is stirred vigorously whilst freezing, ice cream sets unevenly and forms nasty crystals. An ice cream maker is just a double saucepan with a built-in stirrer. In the outer pan goes freezing brine, which you make by dissolving as much salt as the water will take, then freezing it in the fridge ice-tray or your deep-freeze. The advantage of brine is it freezes at a colder temperature than pure water. You then put your pre-cooled ice cream mixture in the inner pan and stir it until it forms a thick creamy genuine ice cream.

COOKING FAT

GHEE is clarified butter as prepared in India for use as cooking oil. It keeps for months, if not years, even in their climate. The purpose of clarification is to separate out the milk-proteins from the butterfat; it is these proteins that make butter go off. What you are left with is almost pure butterfat (plus salt if you added it in the first place).

The basic way to make ghee is to cook butter on a slow heat. and allow all the water to boil out. At the point when the sediment begins to brown, take it off the heat. Pour off the clear liquid through muslin and store it in closed jars. The trouble is that if you leave it just a little too long, the sediment burns and gives a nasty taste to the ghee.

Here is a safer method. Heat the butter on a medium heat until the froth has all been stirred in. Cool the butter and keep it in a fridge for three to four hours. You will see that the ghee sets in a thick layer on the top with a protein sediment underneath. Take off the ghee carefully with a knife and boil it up again for one to two minutes. This produces crystal-clear ghee.

USES FOR WHEY

WHEY CHEESE. Carrying on the principle of not wasting

81

any of that good milk, here is a Norwegian recipe for using up the whey from cheese making. Boil it slowly until it has evaporated to a creamy consistency. Then stir it well and keep boiling until it is really pastelike. Spoon this into a greased bowl to cool. Then tip it out onto a plate for serving. It is brown, highly nutritious and much sought after in Norway. You may find that it takes a bit more getting used to.

BLAAND (from "Recipes of Scotland" by F. Marian McNeill, published by Albyn). "Pour the whey into an oak cask and leave it till it reaches the fermenting, sparkling stage." (Maybe those of us without oak casks could get away with using a plastic wine making bucket?) "It is delicious and sparkles like champagne. After a while, it goes flat - but keep it at perfection by regular addition of fresh whey. Blaand used to be in common use in every Shetland cottage."

A COUPLE OF EASTERN ODDBALLS

KHOA. This dried whole fresh milk is the basic ingredient of Indian milk sweets. Boil a pint of good creamy milk in a thick-bottomed pan on a fast heat. Stir gently in the initial stages - just to prevent it from boiling over. Then, when it starts to thicken, stir more vigorously to keep it from burning. Eventually, in about twenty to twenty-five minutes, you end up with a single lump in the bottom of the pan. That is Khoa. Remove the pan from the heat, but keep stirring till it stops sizzling.

KEFIR. This is a rather peculiar fermented milk, which is alcoholic and gassy. You may have to get hold of Kefir grains, also known as 'yogurt making plants' which go white and gelatinous in milk. Use skim milk at about sixty degrees Fahrenheit and shake the grains and milk from time to time. The grains will have done their work after about twelve hours. They can be drained off, dried and stored. If you cork the milk milk for another twelve hours, you get Kefir. Strange stuff, but popular amongst the wandering tribes of Central Asia.

82

Longford Factory (Interior).

REFERENCES

1. Family Expenditure Survey, HMSO, 1973.

2. Kirschgesner, Nutrition and the Composition of Milk, Crosby Lockwood.

3. Advances in Cheese Technology, FAO, 1958.

4. D. Mackensie, Goat Husbandry, Faber and Faber.

5. If you want to get hold of some goat's milk either to try it for flavour, or to help you with some ailment, Mr. and Mrs. B. Franche will be happy to supply you with dried goat's milk through the post. As far as we are aware, they are the only people producing dried goat's milk in this country. Their address is: Orchard House, Hurst, Martock, Somerset.

6. J.G. Davis, Cheese; Volume 1 - Basic Technology, J and A. Churchill, 1965.

7. Rations for Livestock, Min. of Agriculture (leaflet).

8. Newman Turner, Herdsmanship, Faber and Faber.

9. K. Russell, Principles of Dairy Farming, Illiffe.

10. Richmond's Dairy Chemistry, Griffen, 1953.

11. Mrs. Balbir Singh, Indian Cookery, Mills and Boon, 1961.

12. John Ehle. The Cheeses and Wines of England and France, Harper and Row, 1972.

13. F. Marian McNeill, Recipes of Scotland, Albyn.

MORE HELPFUL BOOKS

Val Cheke: The Story of Cheese Making in Britain, Rout-
ledge, 1959. A very readable account of the amazing variety
of, and interest in, cheese making since Roman times.
Plenty of useful bits on the way for the home cheese maker.

Judkins: Milk Production and Processing, John Wiley, 1960.
An American book for the commercial dairy farmer - well
illustrated and informative about 'modern advances'.

Elspeth Huxley: Brave New Victuals, Chatto and Windus,
1965. A frightening survey of modern factory farming and
agribusiness methods and also, what it could be doing to
our insides.

D. Hartley: Food in England, Macdonald, 1954. 675 pages
of fascinating snippets from this lady's delving into old
books on the eating habits of our ancestors. We have plenty
to learn from them.

How to get back to the Land, The Mother Earth News Special Issue, 1970. This is a reprint of a pre-war homesteading
plan with introductory sections on just about every aspect.
Sketchy and dated, but plenty of good ideas.

Acidophilus and Bifidis, an article in 'The Last Supplement
to the Whole Earth Catalogue', 1971.

Tisdale and Robinson: Buttermaking on the Farm and at the
Creamery,John North 1906. They don't do books like this
any more - we learned a lot from this one.

Sheldon's Dairy Farming, Cassell 1876 - an enormous,
leather-bound beauty that one of us found in a junk shop and
has been referring to ever since. We have used it for a lot
of our illustrations.

John Seymour: <u>The Fat of the Land,</u> Faber ,1961. How a traveller/broadcaster turned homesteader in a remote area of Suffolk. There is an amusing and informative chapter called 'We buy a Cow'.

<u>All About Yogurt</u>, Milk Marketing Board 1972 - a free leaflet all about the stuff the factories make - pasteurised, homogenised, and otherwise improved to give it a shelf life of two weeks.

William Cobbett: <u>Cottage Economy</u>, 1850. This is the original self - sufficiency manual. A delight to read.

K. Russell: <u>Principles of Dairy Farming,</u> recommended by Richard and Joan Collins of Westbury, Somerset. For the more advanced backyarder.

John and Sally Seymour: <u>Self-Sufficiency</u>, Faber and Faber, 1973 - most informative and comprehensive.

Ken Kern: <u>Owner-Built Homestead</u> , Owner-Builder Publications, 1974 - very good and imaginative approach on establishing a self-sufficient homestead - rational and scientific where 'Self-Sufficiency' is amusing and absorbing.

<u>Dairy Work for Goatkeepers,</u> British Goat Society, 1973. For 10p, you get lots of details and recipes on butter, cream, cheeses etc..

<u>Dexter Cattle</u>, The Dexter Cattle Society - a free leaflet with a little more detail about this breed.

THE END